The Abney Level

Handbook

How to Use the Topographic Abney Hand Level / Clinometer Tool – A Guide for the Experienced and Beginners, Complete with Diagrams & Charts

By Hartley Amasa Calkins

and Y. B. Yule

Assistant Engineers, Forest Service, United States Department of Agriculture

Published by Pantianos Classics

ISBN-13: 978-1-78987-050-3

First published in 1935

Contents

Chapter One - Introduction

A Handbook for Both Beginners and Experienced Men

During recent years there has been a large increase both in the number of Abney hand levels [1] used by the Forest Service and the activities in which they are needed. A majority of these instruments, probably about 95 percent, are used by men untrained and inexperienced in handling surveying equipment. This handbook is written primarily, but not solely, for the benefit of such men. It should serve all Abney users, for while some portions may be too advanced for beginners and other portions too elementary for experienced instrument men each user may select the information that satisfies his particular needs.

Importance of the Abney

Next to the compass the Abney is probably the most essential and popular surveying instrument in use in the Forest Service. The numerous activities to which it is adapted, its stability and simplicity of design, its ease and speed of manipulation, and the relatively high precision that is so readily obtained by the careful and experienced operator have made this instrument a favorite of all practical field men who have had an opportunity to give it a thorough trial.

In the Forest Service there are probably 25 Abneys in use to one transit or level. This fact alone is a strong argument in favor of the Abney. Too many beginners are apt to regard the transit or level as the all-important surveying instrument and underestimate the value of the Abney. This is natural enough, for practically all surveying schools and engineering handbooks feature the instruments of higher precision.

Value of Experience

It should be realized that this handbook does not cover all of the minutest details and short cuts that the field man obtains through practical experience on the ground. Every experienced man has his own way of handling the minor operations of field activities so as to save time. As the beginner acquires field experience the many minor uses and short cuts will become self-evident. The importance of working out problems on the ground and thus obtaining practical field experience in all operations involving the use of the Abney cannot be overestimated.

[1] *The Abney is also called a clinometer.*

Chapter Two - Design and Description Design

The original design of the Forest Service Abney level was made in 1914 for the purpose of obtaining a larger and more substantial instrument. Subsequently, improvements were made from time to time to meet field needs. Unless otherwise specified, all information in this handbook relates to the improved instrument as illustrated by figure 1. [1]

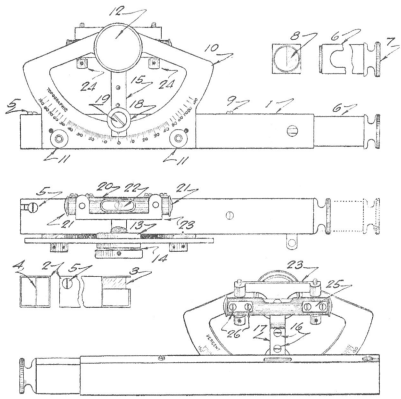

FIGURE 1

1. Telescope tube. 2. Prism and cross-hair slide. 3. Prism. 4. Cross-hair slide.
5. Prism slide lock screw. 6. Slide tube. 7. Eyepiece cap. 8. Half lens. 9. Slide tube lock screw. 10. Graduated limb. 11. Limb capstan nuts. 12. Main capstan bolt.
13. Main clamp nut. 14. Spring washer. 15. Index arm. 16. Index arm screws. 17. Index arm lock bar. 18. Index arm lock nut. 19. Lock nut screw. 20. Vial tube.
21. Vial tube ends. 22. Glass vial. 23. Vial bracket. 24. Capstan adjusting screws.
25. Capstan screw anchor. 26. Capstan screw anchor screws.

Description

The eyepiece slide tube can be pulled out or pushed in, in order to allow for focal adjustment of the individual operator. The index arm-locking device is for the purpose of setting at some definite graduation and is valuable for such operations as surveying trails on a uniform grade. Other improvements are so obvious when the instrument is at hand that they need no additional description.

The Graduated Limbs [2]

There are three types of graduated limbs: Percent, degree, and topographic. The percent limb is based on an angular unit represented by the ratio of 1 unit vertically to 100 units horizontally. The degree limb is based on an angular unit of 1 degree or one three hundred and sixtieth part of a circle. The topographic limb is based on an angular unit represented by the ratio of 1 unit vertically to 66 units horizontally.

On the improved Abney the limb is graduated with the percent graduations on one side and the topographic graduations on the other. Any combination of graduations desired on the limb may be obtained from such firms as A. Leitz & Co., Keuffel & Esser Co., and Eugene Dietzgen Co. when the instrument is purchased.

Designs of Topographic Tapes

The topographic limb and the topographic tape were designed to be used together and the fundamental principles of both are based on "feet rise in 1 chain." Since, therefore, a definite relation exists between a horizontal distance of 1 chain and the topographic limb, a unit length of 1 chain is the basis for designing all topographic tapes. The correction graduations are illustrated in detail by figure 2, A. They are for the purpose of reducing slope distances to horizontal distances as described in detail under (1) Unit Chain Lengths, chapter 4, and illustrated by figure 14.

In accordance with the foregoing principles, three topographic tapes, 2,3, and 5 chains in length, have been designed. Figure 2 is a copy of the original design of the 2-chain trailer tape [3] and, as shown by A, the graduation corrections for 1 chain are placed on the back of the tape back of the 1-chain point. Correction graduations for 2 chains, illustrated by B, are placed on the trailer back of the 2-chain point.

The 3and 5-chain topographic tapes [3] are not trailer tapes. Standard graduations are placed on the front of each tape. Correction graduations are placed on the back of each tape. On the 3-chain tape the graduations are placed back of the 50-link, 1-chain, and 2-chain points; on the 5-tape graduations are placed back of the 50-link, 1-chain, 2-chain, 3-chain, and 4-chain points.

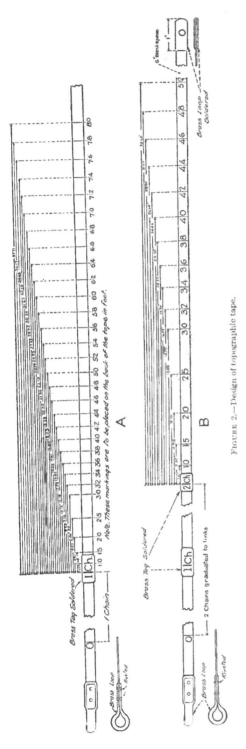

FIGURE 2.—Design of topographic tape.

The designs of the 3 and 5-chain tapes were made recently and the correction graduations were placed sufficiently close together so that a higher grade of precision may be obtained without interpolation than was possible with the older 2-chain tape which was designed for use where speed and not a high degree of precision was desired.

[1] The index arm-locking device (see nos. 17, 18, and 19) is the most recent improvement and will not be found on a large number of instruments now in use. The names of parts were obtained from instrument manufacturing firms and should be used when ordering extra parts.

[2] For purposes of identification the name "percent", "degree", or "topographic" is etched in the upper left corner of each limb. The words "rise in feet per chain" are also etched in the upper right corner of most topographic limbs.

[3] These tapes are now carried as standard stock equipment by such firms as Keuflel & Esser Co., Eugene Dietzgen Co., and Lufkin Rule Co.

Chapter Three - Adjustments

There are three major adjustments to this instrument: (1) Prism and cross-hair slide, (2) height of glass vial, and (3) glass vial parallel to line of sight.

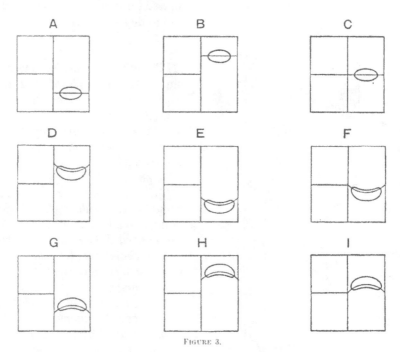

FIGURE 3.

Prism and Cross-Hair Slide

This adjustment is for the purpose of bringing the etched line across the glass vial into coincidence with the cross hair for level readings. This is accomplished by first setting the index arm at 0 on the limb and then moving the slide in or out in the front end of the telescope tube until the image of the black etched line across the glass vial coincident with the cross hair. The positions of the bubble, cross hair and etched lines will appear as shown in figure 3, A, B, and C.

After proper adjustment has been made, a fine line should be scratched, either on the slide or the inside of the telescope tube, in order that the slide may be placed back in proper position after cleaning the prism. This adjustment should always be tested after the slide has been taken out.

Height of Glass Vial

The glass vial parallel-to-the-line-of-sight adjustment should be approximately correct before this adjustment is started. This usually can be determined by inspection or may be accomplished by the method used in the adjustment of the carpenter's level; that is, with the index arm reading zero on the limb block up one end until the instrument is level, turn end for end and take up one-half the error by one adjusting screw. The height of glass vial adjustment is for the purpose of making the cross hair and etched line coincide for all readings. Figure 3, D to I, shows_ the relative positions of etched line, bubble, and cross hair when the instrument is read on steep slopes.

From figure 3 it will be readily determined whether the position of the glass vial is too high, too low, or in its proper position. The adjustment must be made by the "cut and try" method.

Because of refraction of light through the liquid in the glass vial, the image of the bubble becomes distorted to a crescent shape when readings are taken on steep slopes. This distortion can be partly, but not wholly, eliminated by adjustment. Therefore, while using the instrument it is necessary that the image of the etched line intersect the bubble near the flat or concave side. The steeper the angle read, the nearer the etched line and concave side of the bubble should come to coinciding and vice versa. This situation may be studied by placing the instrument carefully in a vise or clamp so that the etched line bisects the bubble. Then look through the telescope tube and note the position of etched line and bubble. Repeat this operation on a number of angles ranging from 0° to 45° and study results.

Glass Vial Parallel to Line of Sight

This adjustment can be made in several different ways, but the two following methods have been found to be the most practical in the field:

Two-Man Method

Set an Abney [1] on a stake about 5 feet high midway between two trees, telephone poles, or other substantial objects, 100 to 200 feet apart, as shown in figure 4, A. The ground should be about level. Level the Abney and establish a level line *ab* between the two objects. The instrument need not be in exact adjustment to produce a level line so long as the set-up is made equidistant from the objects. The broken lines *ac* and *bc* are sighted with the Abney. Whether or not the instrument is in adjustment the triangles *adc* and *bdc* are identical, because angles at *a* and at *b* are equal and *ad* equals *bd* (Abney set equidistant from objects) and dc is common to the two triangles. Therefore, *ab* is a level line.

If line *ab* is too close to the ground for convenient use, establish an auxiliary level line *ef* by measuring equal distances up from *a* and *b*. If several Abneys are to be adjusted, erect some sort of a support *g* (a 2 by 4 about a foot long with a spike near the top, operated similarly to a button on a barn door, is a handy arrangement). The point at *f* should be sufficiently well marked so that it can be seen through the Abney.

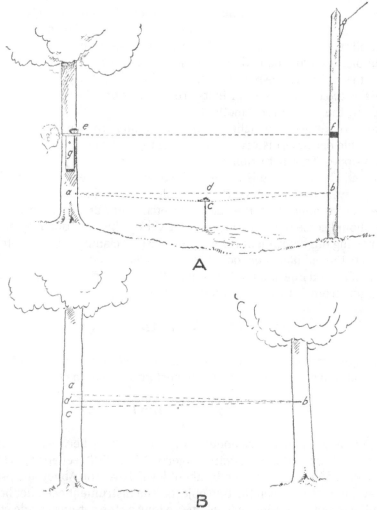

FIGURE 4.—Sketches showing two methods of adjustment.

Set the index arm at zero. Place Abney on support *g* and manipulate *g* until the cross wire is coincident with the sight at *f*. If the bubble does not bisect the etched line, raise or lower one end of the level vial by operating one adjusting screw until the bubble is brought into proper position. When the etched line bisects the bubble and at the same time the line of sight is coincident with the line *ef*, this adjustment is completed.

One-Man Method

Select two trees or other objects about 100 feet apart on nearly level ground, as shown in figure 4, B. Set a mark at *a*; then move to *b*. Set the index arm of the Abney at 0 and sight *a* from *b*; move the Abney up and down at *b* till some point is found which apparently is on a level line through *a*. Mark the point at *b*.

Now move to position *c* and repeat the operations that were performed at *b* and determine point *c*. Set a point *d* midway between *a* and *c* which produces the true level line *db* from which the adjustment should be made in exactly the same manner as described in the preceding paragraph after level line *ef*, figure 4, A, had been established.

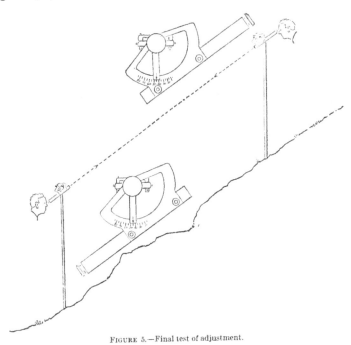

FIGURE 5.—Final test of adjustment.

Final Check of Adjustments

After all adjustments have been made they should be checked to be sure that some adjustment has not been disturbed while some other adjustment was being made. As a final test, read up and down between two definite objects on a steep slope (30° to 45°). If both readings are identical, the instrument is in good adjustment (fig. 5).

[1] When a transit is available It may be used instead of an Abney, if desired.

11

Chapter Four - Uses

Percent Limb

Roads and Trails

The uses of the Abney level in the survey and construction of roads and trails are varied and numerous. A detailed description of each form of use will be attempted here, insofar as the particular use may be desirable in general practice.

This section will be discussed in the following order: Location roads, location trails, and construction.

Location Roads

The use of the Abney on road location is so intimately related to every phase of the work that a detailed description of various duties connected with location will be necessary to bring out properly the value of the Abney in each department of the work. Since, however, this is a treatise on the Abney level, the description is confined to the various uses of this instrument on location and no attempt is made to give a complete analysis of location survey work.

Location surveys may be divided into the following steps:

Preliminary and trial lines.

Center-line location.

Profile.

Cross sections and topography.

Section-line ties.

Preliminary and trial lines. — The importance of the Abney level for running preliminary and trial lines can hardly be overestimated.

Preliminary surveys are more or less elaborate according to circumstances. But whether the survey is to be of sufficient detail to permit the platting of the hne, profile and cross sections from which a projected location may be made, or whether it will consist only of a trial line for the purpose of determining the grade necessary to reach a certain objective, the Abney is rapidly forcing recognition as the most desirable instrument for this work.

The use of this instrument in connection with the more elaborate preliminary surveys will not be discussed here as its uses on location surveys, as described herein, are practically the same.

In mountainous or rolling country it is frequently necessary, over all or part of the route, to employ a certain maximum grade. The work required to determine what grade is necessary is quickly and accurately done with the Abney.

It may be assumed, to illustrate this point, that the route is required to reach a certain pass. The engineer with two or three assistants will start in the pass and run trial lines down on different grades to determine what grade is necessary. This is done easily with one Abney but is greatly speeded up with two. No measurement of distance is required. The Abneys are set on the grade to be used (see note 1, ch 1.). The observation is best taken from the top of a stick cut to a convenient length to the top of a second stick of the same length. See figure 15, A. The length of the sight should be from 100 to 300 feet or longer, depending, on the topography of the country. When the reading is being taken, the flag is motioned uphill or down, as the case may be, until it is in the correct position for the grade being used. The axmen follow through on this line, lightly blazing the line, or, if there are no trees, setting occasional stakes that the position of that particular grade may later be located if desired. Any number of lines may be run in this manner over different grades. Under ordinary conditions in the mountains, from 1 to 2 miles an hour can be run.

Center-line location. — In making the final location of the center line, the line is "spotted" by running the grade line in a manner similar, but with more care and study, to the method used in running trial lines. In general, there are two types of location: (1) The tangent or base-line method and (2) the contour method.

To illustrate these methods, let it be assumed that the route hes on reasonably uniform slopes in typical mountain country.

Tangent or base-line location. — From the initial point, the grade line is run out along the sidehill and marked with "flags" at frequent intervals. The line of these flags will be irregular, but this process may be continued as long as the horizontal position of the flags does not vary greatly from a straight line from the initial point to the last flag. When it becomes apparent that the direction of the line of flags is changing, the center line is established by placing P. I. flags at each end in such a position that the resulting tangent passes through or close to the grade flags. When the position of this line is placed to the satisfaction of the engineer, the intermediate flags are pulled up and the location is continued in a like manner. This results in what is commonly known as the "cut and fill" method.

Contour method.— It frequently happens, especially on low standard roads, that it is necessary to resort to the contour method, m which case the procedure is the same as described above except that no P. I. stakes are set to obtain tangents. The first stakes constitute the line of the road. These stakes are not on the center line but represent the grade line. The resulting alignment is in conformity with the contour of the sidehill.

It is a common practice in running such lines to run it in 100-foot units, setting each stake on grade; shorter distances being used as the need occurs. Distances are measured with a 100-foot tape.

Profile. —The profile is run immediately following the center-line location. If it is done with a regular engineer's or spirit level, the Abney will be of no great assistance except that by clamping the index arm on zero it can be used as a Locke level for spotting turning points and pegging profile through deep draws. However, on small lateral or minor roads where a high degree of precision is not required, m the location, the profile may be run with the Abney. Tins may be done in two ways: (1) By using the Abney as a Locke level [1] and running the levels in horizontal lines, following the same method used with the engineer's level; or (2) by reading the plus or minus grade from one station to the next. Since the latter method is the most practical for Abney use, it is used for illustration below. The following is a sample set of field notes for this method:

Station to station	Grade of profile	Elevation	Remarks
236–237	+6	{ 3,989.0 3,995.0	
237+50	−5	3,902.5	
237–238	0	3,995.0	
238+70	+7.2	4,000.0	
238–239	+5.5	4,000.5	
239–240	+2	4,002.5	
240–241	−3.5	3,999.0	
241–242	−4.5	3,994.5	Dry draw.
242–243	+11	4,005,5	

Profile taken in this manner is sufficiently accurate for roads of inferior standard where it is not important to carry accurate elevations through.

By using this method, from 2 to 4 miles can be run in a day, or better than double the amount that can be run with an engineer's level.

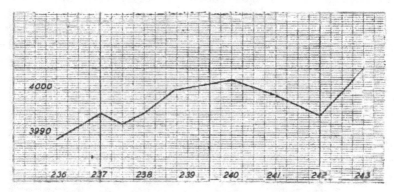

FIGURE 6.—Profile.

Figure 6 represents the platted profile.

Cross sections and topography. — Cross sections and topography may be conveniently worked together, the topography being platted as the cross-section notes are called out.

14

It is not the present practice to sketch topography on location and when done is mostly confined to the higher standard projects. However, since the method described below is one that has been tried and found to be accurate and economical, it is logical that an explanation of its use should be included as representing another use of the Abney level.

The cross-section party consists of a recorder and two rodmen.

When topography is sketched, a topographer is added. The recorder stands on the center line and records the notes as they are called out by the rodman. He also assists these men when needed. Each man is equipped with a rod or stick cut to a convenient length (all must be the same length). Each rodman is equipped with an Abney.

The notes taken are recorded in the following manner:

Station	(Left) L	(Center line) C-L	(Right) R
236	$\frac{-42}{x}$		$\frac{+38}{x}$
237	$\frac{-38}{x}\ \frac{-50}{12}$		$\frac{+40}{x}$
+50	$\frac{-43}{x}$		$\frac{+44}{x}$
238	$\frac{-48}{x}$		$\frac{+49}{10}\ \frac{+46}{x}$
+75	$\frac{-48}{x}$		$\frac{+49}{x}$
239	$\frac{-60}{x}\ \frac{-40}{100}$		$\frac{+42}{50}\ \frac{+52}{30}$
240	$\frac{-55}{x}\ \frac{-42}{50}\ \frac{-32}{35}$		$\frac{+32}{40}\ \frac{+25}{50}\ \frac{+32}{x}$
241	$\frac{-40}{x}$		$\frac{+20}{x}$
242	$\frac{-16}{x}$		$\frac{+16}{x}$
243	$\frac{-30}{x}\ \frac{-10}{100}$		$\frac{+30}{x}$

Using station 240 as an example, the following method is used to take the cross section. One of the requisites of the rodman is a clear understanding of the theory of cross sections. With this in mind, no explanation of the principle involved will be made. The rodmen move out to the right and left of the center line. They are kept in a position at right angles to the center line by the recorder, who remains on the line. The rodmen take position at the first break in the ground line. The percentage of slope is then read by resting the Abney on the stick and sighting to the top of the stick held by the recorder. See figure 15, A. The reading is called out and recorded, together with the distance (horizontal), which is arrived at by measuring with the rod or a tape. As it is hardly ever necessary to take the readings out more than from 50 to 75 feet, any reading that extends up to or beyond these limits may be marked (x) in place of a distance; generally this may be taken to mean 50 feet or more. Figure 7 shows the notes at station 240 platted on a scale of 1 inch = 20 feet. The first reading on the right is +32 percent for 40 feet, the second

+25 percent for 50 feet, etc. For the first reading the rodman sights downhill. If the second reading will go beyond the distance out required, he may simply reverse his position and read the next slope by sighting uphill in a line parallel to the slope. If numerous short readings are required, he must be assisted by the second rodman. Many desirable short-cut methods are worked out by experienced men that permit rapid and accurate work.

At the same time that the cross-section notes are being taken, topography may be sketched. The necessary procedure for this is as follows:

The center-line traverse should be platted on thin paper, the paper to be in a continuous roll about 12 inches wide. Upon this traverse the even contours are located. This is taken from the profile. (See fig. 6.) The elevation of each station and prominent point is also shown. These elevations are marked with small figures inclosed in circles, as shown on figure 8.

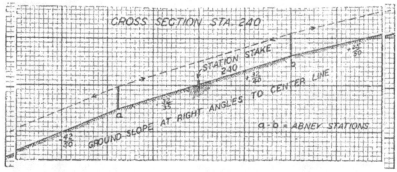

FIGURE 7.—Cross section.

The topographer is equipped with pencil, eraser, 6-inch flat scale, and sketch board. The best type of sketch board is the standard Forest Service traverse board on light tripod. The board is equipped with a tin tube on each end. The paper is rolled from the tube on one side across the board to the opposite tube. The board is light and can easily be carried in one hand and is quickly set up. This method leaves the topographer free to move about when necessary to study certain details of the topography. The sketch board and map in position are illustrated by figure 9.

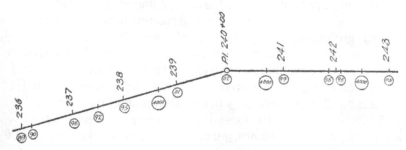

FIGURE 8.—Center line of transverse station numbers—elevation at station points—and intersection of even contours with center line—taken from profile.

16

The topographer works just back of the cross-section party and plats the topography as the notes are called out by that party. Figure 10 shows a section of topography partially completed.

Taking station 240 as an example, the following is the method for platting. The first reading to the right is +32/40. Scale 40 feet to the right and make a dot with a small circle. Thirty-two percent for 40 feet gives a rise of 12.8 feet. Taking the nearest foot would make this 13 feet; added to 02 elevation at the center would make +25 elevation 15 at the 40-foot point. The next reading is +25/30 which would mean a rise of 12 feet. Scale 50 feet from the 40-foot point and mark eleva-

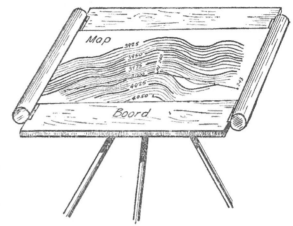

FIGURE 9.—Topographic sketch board.

tion 27. The last reading is +32/X. Scale 100 feet from the last point and also indicate the halfway point. Plus 32 percent for 50 feet means a rise of 16 feet; added to 27 makes an elevation of 43 feet at the halfway point; 32 feet added to 27 makes an elevation of 59 for the last point. The rest is simply a matter of interpolating between these points for position of the contours. The notes on the lower side are platted in the same manner except that the elevation is decreasing instead of increasing.

As previously stated, experienced men will devise many desirable short cuts that will speed up the work without injuring the accuracy or quality of the map. Under average conditions in mountains, timbered country an experienced crew will sketch from 1 to 2 1/2 miles per day, sketching a strip from 150 to 300 feet wide.

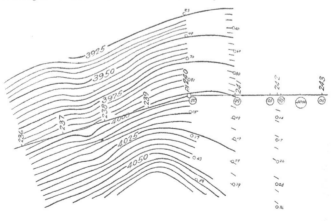

FIGURE 10.—Topography in course of construction.

Figure 11 shows a section of completed map.

Section-line ties. — The Abney performs a valuable service in making ties between road or trail survey lines and section corners. Ties are usually made by running traverses or following section lines. Slope distances are measured with a 100-foot tape and slopes are read with the percent Abney.

The slope distances can later be reduced to horizontal distance by using Table 1, which is both described and shown in chapter 5.

Location Trails

On trail location the Abney is the only instrument required and is far superior to all other instruments. The details of location require no explanation as the world is very similar to that described under preliminary and trial lines for roads, except that the definite location of the trail is staked out in the same manner.

Construction

On road and trail construction the Abney can be used both as a slope Abney and a Locke level [1] for retracing the grade line, setting slope stakes, running out drain ditches in staking out bridges and culverts, and for numerous other purposes.

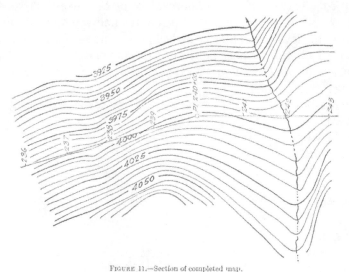

FIGURE 11.—Section of completed map.

Tree Height Measurements

Tree heights are essential for two principal purposes: (1) For determining height growth in connection with growth and yield studies as carried on by forest experiment stations; and (2) to some extent in determining the number of logs per tree, or merchantable height, in connection with timber esti-

18

mating and appraisal activities. The Abney level is the most popular instrument used in this work.

The principles involved in obtaining heights for these two studies are fundamentally the same. The only difference lies in the fact that in growth studies the height of the tree is taken from the ground to the tip of the crown (total height), whereas in timber estimating and appraisal activities the height is measured from the estimated stump height to the "top" diameter (merchantable height). For example: If this top diameter is 6 inches, the tree height is measured from the estimated top of the stump to the point on the tree where its diameter is estimated to be 6 inches.

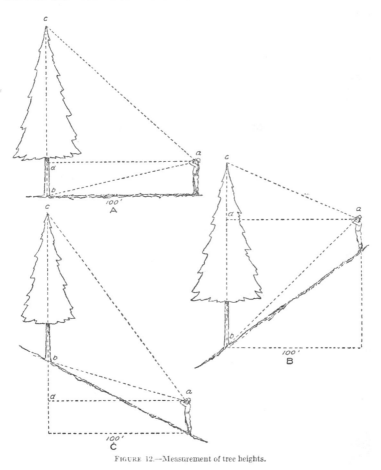

FIGURE 12.—Measurement of tree heights.

Figure 12 illustrates three conditions under which it may be necessary to take measurements. When the percent Abney is used, the horizontal distance (an even 100 feet being preferred) is measured out from the foot of the tree and readings taken as illustrated. If the horizontal distance is 100 feet then in Figure 12, A and B, the sum of the Abney readings, and in C the difference of the Abney readings gives the total height of the tree in feet. If the horizontal distance is more or less than 100 feet the height read must be adjusted accordingly. For example: If the Abney man were only 50 feet or one-half the distance away from the tree, the total height would be one-half the sum of the Abney readings for A and B, and one-half the difference of the readings in C.

When conditions are favorable for growth studies the method illustrated in figure 12, B, is generally used for two reasons: (1) The Abney man has a better chance to see the exact top of the crown; (2) the maximum reading of the Abney is not generally reached. The maximum reading that can be successfully obtained with the percent Abney is about 120. It will be readily seen that if the position of the Abney man in B is such that *bd* equals *dc* it would be possible to measure a tree 240 feet high without increasing the length of base beyond 100 feet. On the other hand, by the method illustrated in C, it would not be possible to measure a tree 120 feet high without increasing the base.

Logging Engineering

The Abney is used extensively by timber appraisers and logging engineers for obtaining grades of streams and slopes to assist in working out railroad, flume, and chute construction costs.

Degree Limb

Tree Height Measurements.

When the heights of a large number of trees are to be measured on steep slopes and in dense canopied forests, as is done by experiment stations in connection with yield studies, the degree Abney has been foimd to be superior in several respects to either the per cent or the topographic Abney. With the degree Abney the observer does not need to be a known horizontal distance from the tree. The instrument (used on a staff) may be set up at any convenient point from which the tip and base of the tree can be clearly seen and slope distances only need be measured. This method materially reduces the time required Tor field measurements. It is illustrated by figure 13, A. The observer sets up at *a* and measures the angles to *b* and *c* and the slope distance *ac*. The triangle *abc* is then solved for side *a*, the height of tree, by means of a slide rule device requiring but one setting. This solution, shown in figure 13, B, is much superior to the long logarithmic solution shown at figure 13, A, and enables the computation of tree heights in the field as rapidly as the measurements can be taken.

The slide rule solution is made as follows:

On scale A find slope distance 90, under this on scale B set the vertical angle M = 50°. On scale C find number equal to total angle M + N = 59°. Then under this on scale D read number 120, which is the height of tree in feet.

The construction of the rule is very simple. The same principles and the same scales are employed that may be found on an ordinary slide rule, but scales are arranged and graduated so as to enable one to do this particular job with ease and dispatch. All scales are in logarithmic terms, the A and C scales being logarithms of numbers and the B and C scales logarithmic sines of angles. If an ordinary 20-inch slide rule is available, logarithmic distances that are accurate to the nearest foot can be scaled off on proper sized pieces of paper and pasted over the ordinary rules of the scale. With skilled workmanship rules of the same size could be made capable of computing heights to the nearest tenth of a foot, a greater degree of accuracy than is usually demanded in field work. For convenience in field use the 20-inch straight rule can be calibrated inside a 7-inch circular rule. This rule is set by rotating one scale inside the other with the same effect as that of moving the slide in a straight rule. This device has been made a part of the Abney by using a complete arc with slide rule graduations.

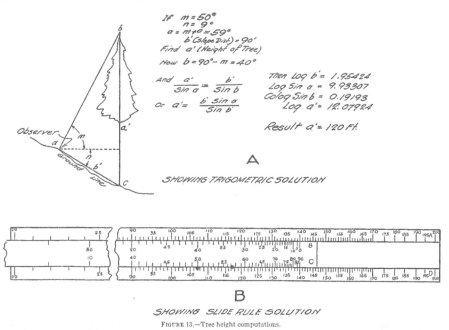

FIGURE 13,—Tree height computations.

The degree Abney and circular slide rule [2] and also the complete arc with slide rule graduations have been given a thorough test at the Northern Rocky Mountain Experiment Station in yield studies requiring the measurement of a large number of tree heights, and found exceedingly satisfactory in field use.

Topographic Limb

Measurement of Lines

Topographic Abney [3] and topographic tape method

Unit chain lengths. — The topographic Abney and the topographic tape, as described in chapter 2, are probably the most satisfactory equipment for fast and accurate Une measurement. By this method a precision sufficiently high to meet practically all Forest Service needs is obtained at a relatively low cost.

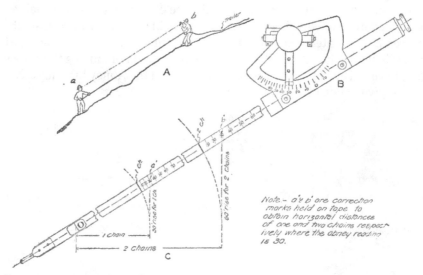

FIGURE 14.—Horizontal and vertical components of slope distances—slope chaining—correction for distance

The method is illustrated by figure 14, A. The positions of two chainmen, *a* and *b* (A), chaining a line down a slope (the method is identical when chaining up a slope) are shown. The Abney is generally carried by chainman *b*, who is at the rear end of the tape, but sometimes, in order to guard against errors, an Abney is carried by both chainmen and notes recorded independently. Suppose that the last pin "stuck" by *a* is at the point occupied by *b* and that *b* has "snubbed" or stopped *a* at the 2-chain mark. (The slope distance between a and 6 is 2 chains.) *b* takes a "shot" or "reading" on *a* at a point that is the same distance from the ground as *b's* eye, which is the height of instrument, or H. I. [4] Suppose that the reading on *b's* Abney is 30 (figs. 14, 13); *b* finds the 30 correction mark beyond the 2-chain brass tag (see *b*, fig. 14, C) ; and a tightens tape and sticks pin at the 0 mark on the front end of the tape. The horizontal distance between *a* and *b* is now 2 chains. (In practically all line measurements it is the "horizontal" and not the "slope" distance that is required.) If it is desired to make a 1-chain instead of a 2chain "measurement," the 30 correction mark beyond the 1-chain brass tag is held and a horizontal distance of 1 chain is obtained. (See *a,* fig. 14, C.) The above procedure can then be summarized as follows: Hold a correction mark back of the 1 or 2, etc., chain mark that corresponds with the reading taken by the Abney; that is, if 30 is read on the Abney, hold 30 on the tape beyond either the 1 or 2, etc., chain mark, depending on whether a horizontal unit length of 1 or more chains is desired. Figure 14, B and C, show the correction held on the tape corresponding to the reading on the Abney. The procedure given above covers the fundamental principles of this method of chaining.

There are numerous details that can only be acquired by practical field experience and are so self-evident when on the ground that only ordinary good judgment is needed in mastering them.

The greatest advantages of this method are: (1) No horizontal correction tables are needed, (2) distances can be recorded in unit lengths of one or more chains, (3) manipulation is mechanical and therefore almost fool proof, and (4) the operations are simple.

Fractional chain lengths. — It sometimes happens that a line cannot be measured to unit chain lengths, such as 1 chain or 2 chains, but must be measured to fractional chain lengths such as 0.87 chain, 1.67 chains, etc. There are two principal circumstances under which this condition happens: (1) Tying into triangulation stations, section passing obstacles such as Figure 15.— Height of instrument. corners, or other established points; (2) small cliffs, knolls, etc.

FIGURE 15.—Height of instrument.

Suppose that chainmen *a* and *b* are measuring a line as illustrated by figure 16, and because of a small ledge or cliff on line' it is impossible for *b* to see *a* at a unit horizontal distance of either 1 chain or 2 chains. Chainman *a* is compelled to select a point at the edge of the cliff which is the only suitable site from which shots may be taken both forward and backward. Suppose also that chainman a "sticks" at this point and chainman *b* reads 1.24 chains on the, tape and 74 ½ on the Abney; b's problem is then to find the correct horizontal distance corresponding to a slope distance of 1.24 chains and an angle 74 ½.

By consulting table 2, it will be found that the horizontal distance corresponding to a slope distance of 1 chain (100 links) and an Abney reading of 74 ½ is, by interpolation, 66.1 links. The horizontal distance corresponding to a slope distance of 24 links and an Abney reading of 74 ½ is by interpolation 16 links. The total horizontal distance corresponding to a slope distance of 1.24 chains and an Abney reading of 74 ½ is, therefore, 66.1 + 16 = 82.1, which is for all practical purposes 0.82 chain.

Platting and recording field notes are much simplified by chaining by unit lengths. For this reason it is advisable, after a measurement involving a fractional chain length has been made, to take an additional fractional measurement that will make up a unit length. For example: In using the table in connection with the above problem, a horizontal distance of 0.82 chain is obtained. It would be advisable to take an additional measurement of 0.18 chain so that the sum of the two measurements would make a unit length of 1 chain. Where only a few links are to be measured and no obstacles interfere, as in the above case where only 0.18 chain is needed to obtain an even chain length, the horizontal distance can easily be measured by "breaking chain." This is done by raising the lower end of the tape until the part of the tape used is level or in a horizontal position. The desired division IS then brought vertically over the point from which the measurement IS to be made by using a plumb line or dropping a small stone.

Topographic Survey

Topographic abney and topographic tape method

Obtaining differences of elevation. — In measuring lines horizontal distances only are considered. The making of a topographic or contour map requires the consideration of vertical distances or elevations as well as horizontal distances. It is only a step from measuring lines or obtaining horizontal distances to obtaining vertical distances or elevations.

Figure 14, B, shows an Abney reading of 30 and, since the topographic plate is graduated in angular units corresponding to a rise of 1 foot vertically to 1 chain horizontally, the reading of 30 represents a rise of 30 feet vertically to 1 chain. When the unit length measured is 2 chains instead of 1, and the difference in elevation for 2 inches is desired, the Abney reading is multiplied by 2. If the unit measured is 2 chains and the Abney reading 30, the difference in elevation is 2 X 30, or 60 feet. (See fig. 14, C.) In like manner, if the horizontal distance is 5 chains and the Abney reading is 30, the difference of elevation is 5X30, or 150 feet. Therefore, to obtain the difference of elevation between two points, multiply the horizontal distance (in chains) between them by the Abney reading.

After a horizontal distance has been obtained either by unit or fractional chain measurements the corresponding difference in elevation is always obtained by multiplying that horizontal distance by the Abney reading. [5]

Importance of accurate chaining. — Reliable elevations are directly dependent on accurate line measurements and standard topographic maps are equally dependent on reliable elevations. Therefore, if accurate topographic maps are to be made by this method it is absolutely essential that horizontal measurements be accurately made. For this reason too much importance cannot be placed on the value of good chaining.

Field notes. — The following tabulation is a very convenient form of recording and handling field notes obtained by the topographic-Abney and topographic-tape method when both elevations and horizontal distances are desired:

Course	Dist.		Slope	Diff.elev.	Elev.	Remarks
N. 10° E.					4, 600	Begin at BM in NE. ¼ sec. 10,
		2	− 6	− 12	4, 588	T. 7 N., R. 11 E., elevation
		2	− 5½	− 11	4, 577	4,600. Desc. NE. slope.
		2	− 12½	− 25	4, 552	
		1	− 15	− 15	4, 537	
		1	− 17	− 17	4, 520	
	10	2	− 20	− 40	4, 480	
		2	− 15½	− 31	4, 449	
		2	− 10½	− 21	4, 428	
		2	− 7	− 14	4, 414	
SS ¹	17	1	− 2	− 2	4, 412	Birch Creek S. 47° E. − 5; N. 50 W. +7.
		2	+5½	+11	4, 425	Asc. south slope.
	20	2	+9½	+19	4, 444	

[1] SS is an abbreviation for "side shot" and is used where the data recorded are detached or independent of continuous line notes.

An SS is always "boxed off" (enclosed by hues) in order to avoid confusion with major notes. For example: If the SS recorded in the above notes were not "boxed off", a total distance of 21 instead of 20 would be obtained, thereby introducing a 1-chain error in the total distance.

The first three columns— "Course", "Dist.", and "Slope"— are the most important because actual field data are recorded thereunder. The columns "Diff. elev." and "Elev." are not so important because the data are derived from columns "Dist. and Slope." For convenience horizontal lines are placed at 10-chain intervals and the accumulated distances indicated to the left in the "Dist." column as shown by the numbers 10 and 20. Under "Remarks" opposite SS the course of Birch Creek is shown to be S. 47° E. and N. 50° W. The figures -5 and +7 are slope readings up and down Birch Creek (described more fully later on).

Figures 17 and 18 are copies of two specimen sets of chainman's field notes and represent a compilation or consolidation of the best features of several methods of keeping notes which are in use by the General Land Office. The cumulative distances in chains and the cumulative differences in elevation make possible an easy and rapid method of transferring information from field notes to field maps. For this reason both sets of notes are especially adapted for use in connection with topographic surveys.

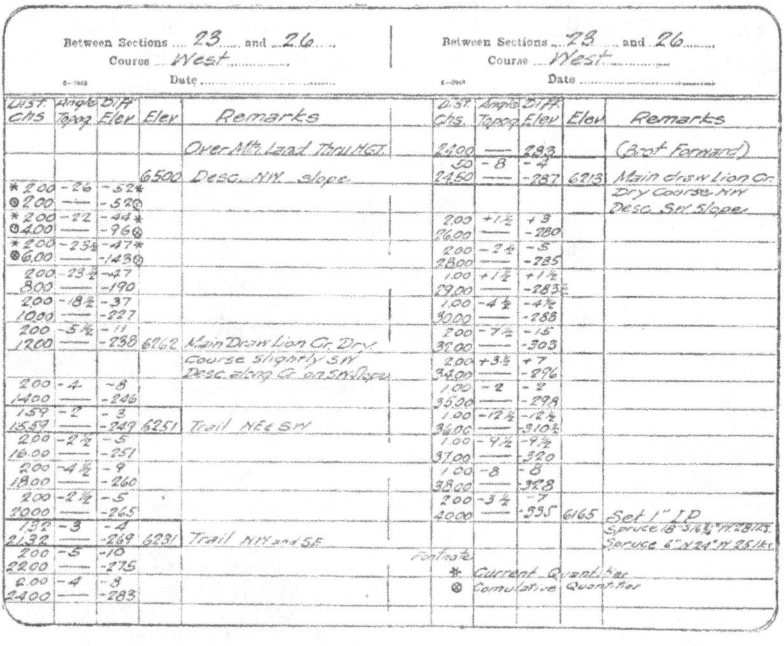

Between Sections 23 and 26
Course West
Date

Dist. Chs	Angle Topog	Diff Elev	Elev	Remarks
				Over Mt. Lead Thru Hgt.
			6500	Desc. NW. slope
*200	-26	-52*		
⊗200	—	-52⊗		
*200	-22	-44*		
⊗400	—	-96⊗		
*200	-23½	-47*		
⊗6.00	—	-143⊗		
200	-23½	-47		
800	—	-190		
200	-18½	-37		
1000	—	-227		
200	-5½	-11		
1200	—	-238	6262	Main Draw Lion Cr. Dry
				Course slightly SW
				Desc. along Cr. on SW Slope
200	-4	-8		
1400	—	-246		
159	-2	-3		
1559	—	-249	6251	Trail NE & SW
200	-2½	-5		
16.00	—	-251		
200	-4½	-9		
1800	—	-260		
200	-2½	-5		
2000	—	-265		
132	-3	-4		
2132	—	-269	6231	Trail NW and SE
200	-5	-10		
2200	—	-275		
2.00	-4	-8		
2400	—	-283		

Between Sections 23 and 26
Course West
Date

Dist. Chs.	Angle Topog	Diff Elev	Elev	Remarks
2400	—	283		(Brot Forward)
50	-8	-4		
2450	—	-287	6213	Main draw Lion Cr.
				Dry Course NW
				Desc. SW Slope
2.00	+1½	+3		
26.00	—	-280		
2.00	-2½	-5		
2800	—	-285		
1.00	+1½	+1½		
29.00	—	-283½		
1.00	-4½	-4½		
30.00	—	-288		
200	-7½	-15		
32.00	—	-303		
2.00	+3½	+7		
34.00	—	-296		
1.00	-2	-2		
35.00	—	-298		
1.00	-12½	-12½		
36.00	—	-310½		
1.00	-9½	-9½		
37.00	—	-320		
1.00	-8	-8		
38.00	—	-328		
200	-3½	-7		
4000	—	-335	6165	Set 1" I.P.
				Spruce 18" S 16½° W 28 lks.
				Spruce 6" N 24° W 25 lks.

Footnote
* Current Quantities
⊗ Cumulative Quantities

FIGURE 17.—Specimen field notes.

Transfer of field data. — A topographic survey is naturally divided into two operations: (1) The control system or foundation from which the map is made, and (2) filling in interior map detail or sketching topography. Lines that are chained and upon which elevations are obtained as described above become a part of the foundation or control system of a topographic survey. The control system, to be of the greatest value, should, when complete, be a "skeleton" topographic map; that is, a complete topographic map should be made along control lines or around control points. The transferring of data from field notes to field map sheets is, in reality from a topographic standpoint, constructing the skeleton topographic map.

To illustrate and clarify this operation, the following description is given of the transfer of data from figure 17 to figure 19 on a scale of 2 inches = 1 mile and a 100-foot contour interval. (Figure 19, form 493, is a special map sheet used by the Forest Service for field use. The dotted lines are convenient for platting lines run in the field in cardinal directions inasmuch as the dots are a uniform distance apart and eliminate the scaling of distances. On a scale of 2 inches = 1 mile the distance between dots represents a distance of 2 chains).

Between Sections *23* and *26* Course *West* Date _____   Between Sections _____ and _____ Course _____ Date _____

DISTANCE CHAINS Total	Item	ANGLE TO POG	DIFF ELEVATION +	−	ELEV Total	ELEV	REMARKS
						6500	Over Mtn. Land
2	2	−26		52	−52		Desc. NW slope
4	2	−22		44	−96		
6	2	−23½		47	−143		
8	2	−23½		47	−190		
10	2	−18¾		37	−227		
12	2	−5½		11	−238	6262	Main draw Lion Cr. Dry. Course slightly SW.
14	2	−4		8	−246		Desc. along Cr. on SW slope.
1559	1.59	−2		3	−249	6251	Trail NE and SW
16	2	−2½		5	−251		
18	2	−4½		9	−260		
20	2	−2½		5	−265		
2/32	1.32	−3		4	−269	6231	Trail NW and SE
22	2	−5		10	−275		
24	2	−4		8	−283		
2450	.50	−8		4	−287	6213	Main draw Lion Cr. Dry, course NW.
26	2	+1½	3		−280		Desc. SW slope.
28	2	−2½		5	−285		
29	1	+1½	1½		−283½		
30	1	−4½		4½	−288	*	
	Total		4½	292½			Note – A check of elevations is obtained.
				4½			
				288	*		

FIGURE 18.—Specimen field notes

From the notes (fig. 17) the line is running west between sections 23 and 26, descending over a northwest slope, and the elevation at the section corner is 6,500. The first step is to plat all topographic and cultural features; therefore, the two draws of Lion Creek and two trail intersections are placed at proper distance and in correct direction. The next step is to supply elevations for corners and important intersections in order to provide control points for filling in interior detail which will be described later on; therefore elevations are shown at one-fourth corner, second corner, and the two draws of Lion Creek, All that remains to be placed on the map are the contours. (Platting the intersections of topographic features, such as streams, ridges, etc., first, aids later in placing contours in their correct position.) Since the line is descending over a northwest slope, the first contour will take a northeast and southwest direction and will pass through the section corner, since the elevation of the section corner is 6,500 feet. The notes show that at 4

27

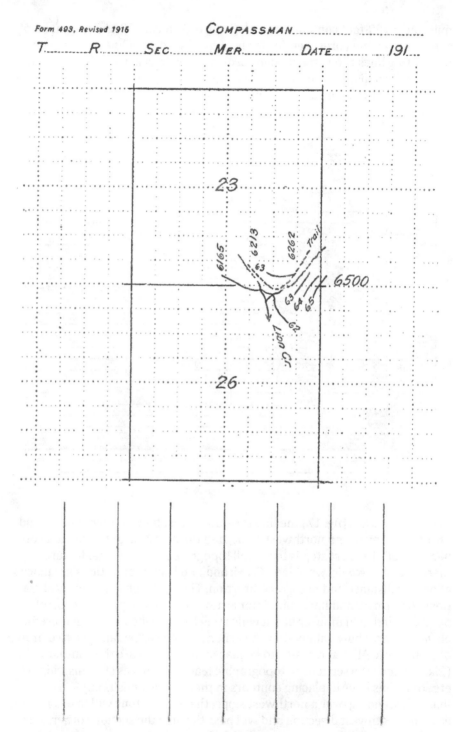

FIGURE 19.—Specimen map platted on form 493 from notes on figure 17.

chains the difference of elevation is 96 feet; therefore, since the contour interval is 100 feet, the 64 (abbreviation for 6,400) contour will be platted in at 4 chains (this is sufficiently close to be consistent with the scale of the map and precision of the method) and will also take a northeast and south, west direction. At 8 chains in the notes the difference of elevation is 190 feet; therefore the 63 contour should cross in a northeast and southwest direction slightly beyond the 8-chain mark (fourth dot) on the map. In like manner, remaining contours are placed according to direction of slope and distance in notes at which multiples of 100 (or nearly so) are found in the "Diff. elev." column. The elevation of 6,213 at the intersection of the main draw of Lion Creek is only 13 feet above the 62 contour, which is sufficiently close to place the 62 contour on Lion Creek.

It is really preferable to handle the above-described operation in the field as it is always easier to sketch the topography on the ground where the terrain can be visualized. However, the notes can be worked up in the office, if absolutely necessary. Whether worked in the field or office, the method of transfer is the same.

For control purposes. — Figure 20 shows an entire section which has been inclosed by transfer as described above. It is in reality a skeleton map or control unit, as has been previously mentioned. It is advisable, when a line is run for control purposes, to have each chainman carry an Abney and record notes independently. One set of notes is for permanent record and the other temporary, or "check notes." Chainmen should compare notes in the field at intervals of from 10 to 20 chains so that in the event of large discrepancies they" may go back on the line until such discrepancies are located while in that locality. By using two Abneys the precision of the survey may be materially raised. Personal errors, such as recording plus for minus angles, are found when notes are checked. If either instrument is out of adjustment the fact will soon show up in elevations and slope readings. It is very seldom that two instruments are out of adjustment at the same time if properly adjusted at the beginning. The use of two Abneys eliminates a lot of lost time and expense that often would be necessary if only one instrument were used and lines had to be rechained in order to obtain desired precision. If one Abney is out of adjustment, the other can nearly always be relied upon for satisfactory notes. When a circuit is closed a better error of vertical closure is always obtained when the average of the instruments is used.

Figure 21 represents a control unit of irregular shape. The methods of line measurement and transfer of data are identical with methods described above where control lines surround a section. Traverse stations instead of section corners are used for control points.

For supplying map detail. — On intensive surveys (usually on a scale of 4 inches = 1 mile) it is often necessary to obtain map detail or interior topography by the Abney and tape method. Usually lines are run in cardinal directions at regular intervals of 10 or 20 chains apart and have become known as

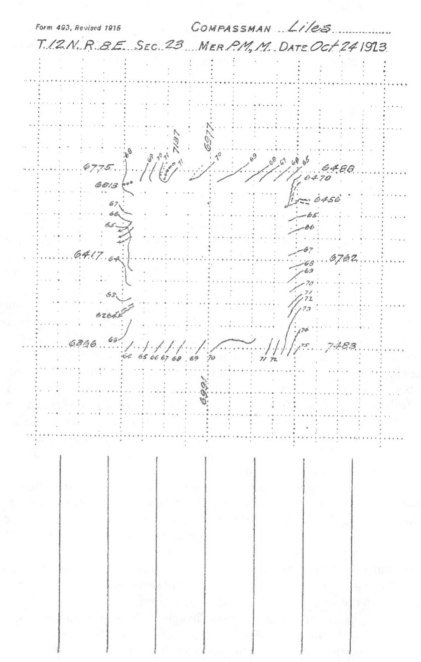

Form 493, Revised 1916 COMPASSMAN *Liles*

T.*12*.N. R.*8E*. SEC. *23* MER.*P.M., M.* DATE *Oct. 24 1913*.

FIGURE 20.—Specimen skeleton map showing amount of data available from G. L. O. chainmen's notebook

"strip lines." The "gridiron," as differentiated from the "traverse" or " trian-gulation " methods, is the name given to an area that has been worked by the strip-line method. Figure 22, A, represents a control unit of four sections. The control around the exterior boundary was established by the Abney and tape

30

method on which two Abneys were used, as has been previously described. Strip lines A to P are also shown and the directions in which they were run indicated by arrows. When the control lines were run, stations were set at regular intervals from which strip lines were initiated and upon which strip lines were terminated.

Strip lines are run by the same method as control lines, only not to such a high precision. Only one Abney is used, on strip lines and it is unnecessary to be so careful in allowing horizontal corrections of slope distances while chaining. Figure 22, B, is a completed topographic map of control unit represented by figure 22, A.

Typographic Abney and Pacing Method

This method is almost identical with the Abney and tape method so far as the instrumental part of the operation is concerned. Distances are obtained by pacing and, instead of running lines in cardinal directions, traverses usually follow "master lines" or lines of least resistance so that good pacing conditions may be obtained. Since the entire operation is done by one man, Abney readings are taken ahead on trees or other objects that are on line and the H. I. (height of instrument) estimated. On figure 23 the stream traverse was run by the Abney and pacing method and shows one use of the Abney in filling in interior map detail.

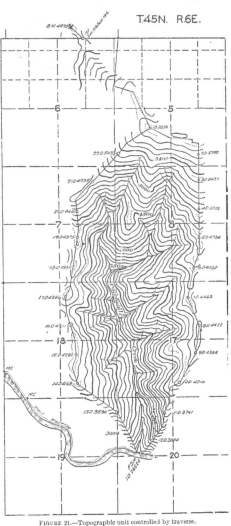

FIGURE 21.—Topographic unit controlled by traverse.

Topography by Abney Alone

On adjacent areas. — As has been explained, the graduations on the topographic plate are based on "feet rise per chain." If a reading is taken up or down a slope, ridge, stream, or other topographic feature, the reading indi-

31

cates the number of feet rise vertically for each chain along the line of sight if the slope is uniform. Suppose a "shot" is taken up a uniform slope or ridge and the reading is -f 25. It is evident that it will be 4 chains (horizontally) between contours if a 100-foot contour is used in sketching topography. Therefore, by scaling from the point where the shot is taken and in the direction it is taken, the placing of contours becomes a mechanical operation. It should be remembered that the contours are spaced up without having to climb the slope. This whole operation is performed from one point. A number of shots by the "Abney alone" method are shown on figure 23 by broken lines with direction arrows.

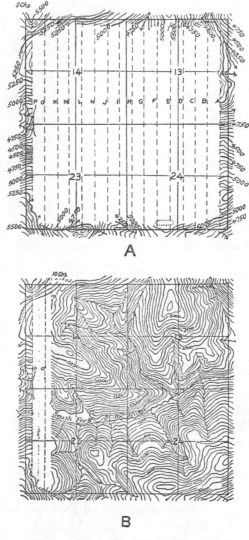

FIG. 22.—Strip or gridiron method

32

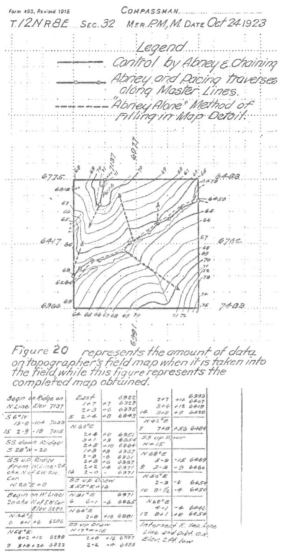

Legend

Control by Abney & Chaining

Abney and Pacing traverses along Master Lines.

"Abney Alone" Method of Fitting in Map Detail.

Figure 20 represents the amount of data on topographer's field map when it is taken into the field, while this figure represents the completed map obtained.

FIG. 23.

This method is shown in detail by figure 24, A, B, and C. Suppose that the topographer is occupying station a, which is on the top of a hill and from which ridges and draws or streams are visible in all directions. Suppose also that the directions of lines have been determined and platted on the topographer's field sheet as illustrated by lines 1, 2, 3, 4, 5, 6, 7, and 8, A. The topographer's problem is then, to build a topographic map of the area around him with the "Abney alone" (tables are not even needed in this operation). The first part of the operation is to take-Abney readings down all streams, draws, and ridges, which in this case are "master lines." The readings, — 5,

— 20, etc., are jotted down along the master lines. The second step is to determine the horizontal distance between contours and place a dot on the master line where each contour crosses. The third step is to space contours along master lines as shown by the solid portions of contours in B. The "stubbed off" contours should show shapes as near as it is possible for the topographer to "sketch" or estimate by eye. For example: The draw on line 4 is much deeper and steeper than draw shown by line 8. After contours have been spaced along master lines, the remaining portions of contours (shown by dotted lines) are very easily sketched in. In fact, there is but little that these portions of the contour can do but "fall" into place. From a study of B, it is obvious why ridges and streams have been named "master lines." The fourth and last step of the operation is to fill in contours between master lines, sketch in topographic and cultural features, such as streams, draws, trails, etc., and the finished map, C, is obtained.

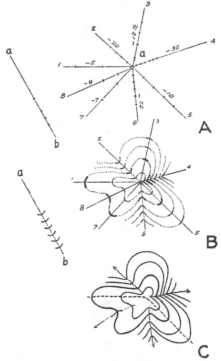

FIGURE 24.—Detailed sketches from one point.

In describing the above method of "spacing" contours as illustrated by figures 23 and 24, it was assumed that the point from which the Abney shots were taken had an elevation "on the contour"; that is, to an even 100 feet as 6,400 or 6,500, but it seldom happens that the elevation of a point is an exact multiple of 100.

Let us suppose the topographer is on a known point on a ridge the elevation of which is 4,626 feet (fig. 25, A) and that he takes a shot of +11 up the ridge and a shot of —27 down. If a 100-foot contour interval is used it is at once evident that the distance is approximately 6 ½ chains up the ridge to the first contour, and contours thereafter are spaced at 9-chain intervals, so long as the slope of the ridge remains the same. In like manner it will be 1 chain down the ridge to the first contour, and contours thereafter will be spaced at intervals of about 4 chains. To expedite operations, calculations incident to spacing contours should be made mentally. To facilitate calculations Abney shots on steep slopes and elevations can generally be thrown to the nearest 5 feet without affecting the accuracy of the map. For example: If the elevation given above were 4,625 instead of 4,626 and the shot — 25 instead of —27 were used, the resulting difference in spacing contours on a scale of 2 inches or 4 inches =1 mile would be negligible because contours spaced in from the two sets of figures would be coincident or nearly so. At any rate, the difference would be too small to be considered. On the other hand, if a shot of +10 instead of +11 were used, it is readily seen that contours spaced at 9 and 10 chains would give a difference of 1 chain, and an error of 1 chain is easily detected on large-scale maps. A little experience will teach the beginner which shots can be approximated and which must be read to the nearest foot and not make an appreciable difference on the scale of the map used. The two following rules may be used in spacing contours when the elevation of the point from which spacing is done falls between contours: (1) Subtract the elevation of the point from the elevation of the next higher contour and divide the difference by the Abney reading for up-grade shots. The result will be the distance to the first contour above the point. (2) Subtract the elevation of the first contour below the point from the elevation of the point and divide the difference by the Abney reading of down-grade shots. The result obtained will be the distance to the first contour below the point.

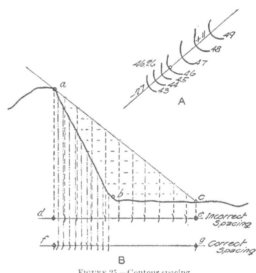

Figure 25.—Contour spacing.

It should be remembered that Abney shots must be taken *along slopes* or the two methods described above cannot be used. Figure 25, B, is designed to illustrate this point. Suppose *a* is the top of a hill; *b* the foot of the slope; *c* is a known point on a flat at which the topographer is located, and it is desired to space contours up the slope *a-b*. The topographer should first establish the position of *b* (both vertically and horizontally) relative to *c* and then from *b* take a shot up the slope *a-b* and space in contours. A shot should never be taken from *c* to *a* on the ground and contours spaced in on the topographer's map from *c* to *a*. The map thus obtained would show that *a* was the top of the hill and *c* the foot of the slope, instead of *a* being the top of the hill and *b* the foot of the slope as actually exists on the ground. It may appear that this illustration is so self-evident that such errors would not be made; yet it is known that such errors are common with amateur mappers. The correct and incorrect spacing of contours as they would appear on the topographer's map sheet are illustrated by lines *de* and *fg*.

On distant areas. — In sketching topography it is often convenient, and in some cases necessary, to obtain a slope reading on some distant ridge too far away from the topographer to obtain a direct reading up or down the slope, as described above. This can be done very easily by bringing the bottom of the Abney telescope tube into coincidence with the profile of the ridge and leveling the glass vial while the Abney is in this position. The slope can then be read directly from the graduated limb.

FIGURE 26.—Readings on distant slopes.

This operation is illustrated in detail by figure 26, A. The topographer is shown leveling the Abney when the telescope tube is parallel to the slope. In the diagram B, let *ab* represent the slope desired, cd the Abney, and the line of sight from the topographer. The best results will be obtained when the desired slope is not over 1 mile distant, when *cd* is parallel to *ab* (or nearly so) and when *ef* is perpendicular to both *ab* and *cd*. In C, *ab* and *cd* are parallel, but *ef* is not perpendicular to *ab* and *cd*. Under these conditions the Abney *cd* should never take the position *cd*; that is, perpendicular to the line of sight *ef* but should always be held parallel to the slope *ab*. The great difficulty in performing this operation in the field is knowing when the Abney is parallel with and the line of sight perpendicular to the slope.

This method is very valuable when field operations are properly executed, but there are so many chances for error that the beginner is cautioned to use great care and be skeptical about using results unless they have been checked and lie is satisfied that desired results have been obtained.

Suppose that while the topographer is at station *a* (fig. 24, A), he desires a profile reading of ridge *ab*. The operation is accomplished as described above and as soon as the reading (which is assumed to be 20) is obtained, spacing for contours is shown by dots in exactly the same manner as shown by A. "Stubbed oft'" contours are also shown in the same manner as shown by B. The horizontal position of *a-b* is seldom definitely known at the time the reading is taken, in which case the "stubbed off" contours hold shape and slope until the position of the ridge *a-b* is actually located.

On minor triangulation. — The Abney is often used on topographic surveys as a supplement instrument in connection with planetable work. It is also used to a great extent in obtaining elevations in connection with minor triangulations. Often when a ridge traverse is run it is necessary to triangulate to some point on an adjacent ridge while passing, and the Abney is a very useful instrument in this type of survey.

The details of use as a triangulation instrument are illustrated by figure 27. That the horizontal positions of points *a, b,* and *c* are known (have been located by compass, planetable or some other type of instrument, either by traverse or triangulation) and also the elevations of *a* and *b*. Assume the elevation of a to be 6,500 feet and *b* 6,400 feet, the distance *cb* to be 50 chains and *ac* to be 30 chains and the Abney readings taken from *a* and *b* to *c* + 13 ½ and +10, respectively. The difference of elevation between *a* and *c* is 30 x 13 ½ = 405 feet, which added to the elevation of *a*, which is 6,500 feet, gives an elevation of 6,905 feet for *c*. In like manner the difference of elevation between *b* and *c* is 50 x 10 = 500 feet, which added to the elevation of *b*, which is 6,400 feet, gives an elevation of 6,900 feet for *c*. Thus, by a very easy process the two elevations of 6,900 and 6,905 feet are obtained for the elevation of c. Either elevation would be sufficiently accurate for the method used, but the second elevation is always valuable as a check. From the above description, the following summary is given: When the horizontal position of two points and the elevation of one point are known, multiply the distance in chains between the two points by the Abney reading from one to the other, which will give the difference of elevation. Add or subtract (depending on whether the unknown point is higher or lower than the known point) the difference of elevation to or from the elevation of the known point and the result will give the elevation desired.

Precision

The precision of any instrument is the degree of accuracy that can be obtained by the use of that instrument on some operation or piece of work.

There has been an excellent opportunity of determining the precision of the Abney level on topographic surveys during the past 12 years. It has been found that, with two good instruments well adjusted and operated by experienced and competent field men, circuits for control purposes, ranging from 10 to 15 miles in perimeter, can be held to an average error of about 1 foot per mile. A compilation of 2,000 strip miles run by both experienced and inexperienced men on several projects under various conditions gave an average of 10 feet per mile for the "gridiron" method of obtaining interior map detail. On cooperative surveys between the General Land Office survey and the Forest Service, a record was kept of lines run around 140 sections (about 90,000 acres) which were used as control units, and an average error of 5 ½ feet per mile was obtained. Only one Abney was used and the chainmen were inexperienced in this type of work.

The above figures are general averages and maximum errors are really what are desired and needed. On all topographic surveys a precision is set which controls maximum errors, and there has been no great difficulty encountered in obtaining the required precision by experienced men, so long as the control units were not made too large.

Scope

The scope of the instrument is, generally speaking, the size of the area upon which it can be used and produce satisfactory results. It is obvious that scope or size of area is governed almost entirely by the precision or degree of accuracy with which an activity is performed.

In preparing special instructions which will govern the operations of the surveyor or topographer in the field, the precision of the map desired is always included. Suppose it is desired to make a topographic map of an area of five townships on a scale of 2 inches = 1 mile and a contour interval of 100 feet. Suppose also that a precision of one-half the contour interval is set for the project. This means that all principal or salient topographic features over the entire area must be located within 50 feet of their true elevation.

From a study of the results obtained, as explained under Precision, (above), together with field experience, it seems topography on an area equivalent to half a township can be held to the precision of half the contour interval, or 50 feet, without any great difficulty. Therefore, the five townships would be divided into 9 or 10 units each tied to a control point of higher precision than can be obtained by Abney. In like manner, where one-fourth the contour interval, or 25 feet, is the precision desired, the control unit should not exceed one-fourth of a township. In other words, a control point of higher precision must be supplied for each one-fourth or one-half township, depending on whether one-fourth or one-half the contour interval is set as the desired precision. On the five townships, 20 units would be needed if a precision of one-fourth contour interval is desired.

The scope of the Abney would then be an area of one-half or one-fourth township, depending on precision. In like manner the scope may decrease or increase in area, depending on whether the precision is raised or lowered.

Costs

The Abney, in addition to raising the standard of topographic maps over old methods used by the Forest Service, has also materially reduced costs.

On timber surveys on a scale of 4 inches = 1 mile, worked by the gridiron method, the entire project costs have ranged from 10 to 20 cents per acre, depending on the type of country and the number of strip lines run per mile. In addition to a topographic map, a timber estimate and silvical data are obtained. Probably about one-half of the project costs would properly be chargeable against the topographic map obtained which, as a general average, would be about 7 ½ cents per acre.

Topographic maps on a scale of 2 inches = 1 mile nave been made for 2 to 5 cents per acre, depending on the method used, type of country covered, and precision and amount of detail desired.

About 12 townships have been covered by cooperative surveys by the Forest Service and General Land Office survey on a scale of 2 inches = 1 mile. Interior detail was supplied by the Abney and pacing method supplemented by triangulation and side shots, and cost about 2 ¼ cents per acre. The cost of General Land Office Survey, which also furnished control, was about 10 cents per acre.

Under varying conditions, by different methods, scales, and precision, the cost of topographic detail upon which the Abney has been used has ranged from 1 ¼ to 7 ½ cents per acre.

Tree Height Measurements

The topographic Abney may be used with a 66-foot base in incidentally the same way that the percent Abney is used with the 100foot base, as described under tree-height measurements, page 15 (percent limb). The topographic Abney should never be operated with a 100-foot base, and likewise the percent Abney should never be operated with, a 66-foot base unless equivalents such as are shown by table 4, are used. It is necessary to caution the beginner on this point, as it is known that men with considerable experience have confused the percent and topographic principles. Since there is a tape designed to be used with the topographic Abney, as previously described, and since there is no trailer tape designed for use with the percent Abney, the topographic Abney should be preferable inasmuch as the horizontal distances can readily be measured by using the topographic tape.

Percent, Degree, and Topographic Limbs

Substitute for Locke Level

By locking the index arm at 0 the instrument may be used as a Locke level. There are two ways that this can be done: (1) When two men are available, a rod is used and a line of levels run similar to the methods used in running more precise level lines. (2) When only one man is available the line is run upgrade. From the initial point some object is sighted on the ground on a level with the observer's eye. The observer then moves to the point sighted and repeats the operation. By this step-up method, elevations may rapidly be carried from one point to another over short distances. The observer's H. I. must be known. Suppose this H. I. is 5 ½ feet and the observer has taken 10 points or stations in reaching his terminal point. It is at once evident that the accumulated difference of elevation is 55 feet.

The Locke level method is often used short distances where speed is desired and a high degree of precision is not essential. This use is valuable on intensive large-scale topographic surveys made partially by one man. [5]

It is at once obvious that when the index arm is locked at 0, limb graduations are not used, and it is therefore immaterial what type of graduated limb the instrument is equipped with (percent, degree, or topographic) when substituted for a Locke level.

Minor Functions

As the beginner becomes more experienced, many minor operations will be performed and new uses developed which will be so self-evident that no description of methods are necessary. This will be true regardless of which of the three limbs — the percent, degree, or topographic — is used.

[1] See "Substitute for Locke level," Ch. 4.
[2] Invented and developed simultaneously but independently by I. T. Haig and G. S. Kempff, of the Rocky Mountain Experiment Station.
[3] The improved Abney equipped with a topographic limb is generally called the topographic Abney. See figure 15 for two different methods of obtaining the H. I.
[4] Where short horizontal distances are obtained, by "breaking" chain, an Abney reading should always be taken in order to obtain the corresponding difference in elevation.
[5] Other uses of the Abney as a Locke level on roads and trails are given in Chapter One.

Chapter Five - Tables

Need for Tables

The Abney levels of both the old and the improved types in use by the Forest Service at the present time are made by several instrument manufacturers, and either per cent, degree, or topographic graduations may be marked on the vertical limb. On some instruments the limb is graduated on both sides as described under "graduations,". In like manner a varied assortment of tapes are in use by the Forest Service and consist principally of the band-steel type, of which a portion are standard and the remainder special equipment. The standard lengths are 1, [1] 2, and 8 chains, 50, 100, and 300 feet. The special topographic tapes are 2, 3, and 5 chains in length as described under Designs of Topographic Tapes.

It is very seldom that any one individual will be in possession of all the equipment that is especially adapted to the various uses of the Abney. For example: One forest officer may have only an old-style link chain and degree Abney, another may have only a 100-foot tape and per cent Abney, and still another may have only a trailer tape and topographic Abney. Five tables have been prepared for the purpose of increasing the range of uses of available instruments. Tables 3, 4, and 5 give equivalents between the per cent, degree, and topographic graduations and should provide sufficient information to make possible the utilization of the instruments at hand on any activity. Table 1 is for use with the per cent Abney and 100-foot tape. It furnishes a means of obtaining horizontal distances when the percentage of slope (Abney reading) and the slope distance are known. By interpolation horizontal distances to the nearest one-tenth of a foot may be obtained. If properly handled the table may be used with a longer tape — the 300-foot tape for example. Suppose that a slope distance of 175 feet has been measured and the slope reading is 25 per cent. From the table the horizontal distance for 100 feet on a 25 per cent slope is 97 feet and the horizontal distance for 75 feet on a 25 per cent slope is 72.8 feet. The total horizontal distance is therefore 97 feet +72.8 feet or 169.8 feet. This use of the table is similar to the use of a traverse table.

If the difference in elevation as well as the horizontal distance is desired, it may be obtained by multiplying the per cent of slope by the horizontal distance.

Table 2 is for use with a 1-chain (100-link) tape and topographic Abney. The method of interpolation and use with tapes longer than 1 chain are identical with the methods just previously described under table 1. The horizontal distances obtained by this table are in units of links and should be divided by 100 to convert to chains.

TABLE No. 1.—*Conversion of slope distances to horizontal distances*

Slope distance, feet	Percent																		
	10	15	20	25	30	35	40	45	50	55	60	65	70	75	80	85	90	95	100
2	2.0	2.0	2.0	1.9	1.9	1.9	1.9	1.8	1.8	1.8	1.7	1.7	1.6	1.6	1.6	1.5	1.5	1.5	1.4
4	4.0	4.0	3.9	3.9	3.8	3.8	3.7	3.6	3.6	3.5	3.4	3.4	3.3	3.2	3.1	3.0	3.0	2.9	2.8
6	6.0	5.9	5.9	5.8	5.7	5.7	5.6	5.5	5.4	5.3	5.1	5.0	4.9	4.8	4.7	4.6	4.5	4.4	4.2
8	8.0	7.9	7.8	7.8	7.7	7.6	7.4	7.3	7.2	7.0	6.9	6.7	6.6	6.4	6.2	6.1	5.9	5.8	5.7
10	10.0	9.9	9.8	9.7	9.6	9.4	9.3	9.1	8.9	8.8	8.6	8.4	8.2	8.0	7.8	7.6	7.4	7.3	7.1
12	11.9	11.9	11.8	11.6	11.5	11.3	11.1	10.9	10.7	10.5	10.3	10.1	9.8	9.6	9.4	9.1	8.9	8.7	8.5
14	13.9	13.8	13.7	13.6	13.4	13.2	13.0	12.8	12.5	12.3	12.0	11.7	11.5	11.2	10.9	10.7	10.4	10.2	9.9
16	15.9	15.8	15.7	15.5	15.3	15.1	14.9	14.6	14.3	14.0	13.7	13.4	13.1	12.8	12.5	12.2	11.9	11.6	11.3
18	17.9	17.8	17.7	17.5	17.2	17.0	16.7	16.4	16.1	15.8	15.4	15.1	14.7	14.4	14.1	13.7	13.4	13.1	12.7
20	19.9	19.8	19.6	19.4	19.2	18.9	18.6	18.2	17.9	17.5	17.1	16.8	16.4	16.0	15.6	15.2	14.9	14.5	14.1
22	21.9	21.8	21.6	21.3	21.1	20.8	20.4	20.1	19.7	19.3	18.9	18.4	18.0	17.6	17.2	16.8	16.4	15.9	15.6
24	23.9	23.7	23.5	23.3	23.0	22.7	22.3	21.9	21.5	21.0	20.6	20.1	19.7	19.2	18.7	18.3	17.8	17.4	17.0
26	25.9	25.7	25.5	25.2	24.9	24.5	24.1	23.7	23.3	22.8	22.3	21.8	21.3	20.8	20.3	19.8	19.3	18.8	18.4
28	27.9	27.7	27.5	27.2	26.8	26.4	26.0	25.5	25.0	24.5	24.0	23.5	22.9	22.4	21.9	21.3	20.8	20.3	19.8
30	29.9	29.7	29.4	29.1	28.7	28.3	27.9	27.4	26.8	26.3	25.7	25.2	24.6	24.0	23.4	22.9	22.3	21.7	21.2
32	31.8	31.6	31.4	31.0	30.7	30.2	29.7	29.2	28.6	28.0	27.4	26.8	26.2	25.6	25.0	24.4	23.8	23.2	22.6
34	33.8	33.6	33.3	33.0	32.6	32.1	31.6	31.0	30.4	29.8	29.2	28.5	27.9	27.2	26.5	25.9	25.3	24.6	24.0
36	35.8	35.6	35.3	34.9	34.5	34.0	33.4	32.8	32.2	31.5	30.9	30.2	29.5	28.8	28.1	27.4	26.8	26.1	25.5
38	37.8	37.6	37.3	37.0	36.4	35.9	35.3	34.7	34.0	33.3	32.6	31.9	31.1	30.4	29.7	29.0	28.2	27.5	26.9
40	39.8	39.6	39.2	38.8	38.3	37.8	37.1	36.5	35.8	35.0	34.3	33.5	32.8	32.0	31.2	30.5	29.7	29.0	28.3
42	41.8	41.5	41.2	40.7	40.2	39.6	39.0	38.3	37.6	36.8	36.0	35.2	34.4	33.6	32.8	32.0	31.2	30.4	29.7
44	43.8	43.5	43.1	42.7	42.1	41.5	40.9	40.1	39.4	38.6	37.7	36.9	36.0	35.2	34.4	33.5	32.7	31.9	31.1
46	45.8	45.5	45.1	44.6	44.1	43.4	42.7	41.9	41.1	40.3	39.4	38.6	37.7	36.8	35.9	35.0	34.2	33.3	32.5
48	47.8	47.5	47.1	46.6	46.0	45.3	44.6	43.8	42.9	42.1	41.2	40.2	39.3	38.4	37.5	36.6	35.7	34.8	33.9
50	49.8	49.4	49.0	48.5	47.9	47.2	46.4	45.6	44.7	43.8	42.9	41.9	41.0	40.0	39.0	38.1	37.2	36.2	35.4
52	51.7	51.4	51.0	50.4	49.8	49.1	48.3	47.4	46.5	45.6	44.6	43.6	42.6	41.6	40.6	39.6	38.7	37.7	36.8
54	53.7	53.4	53.0	52.4	51.7	51.0	50.1	49.2	48.3	47.3	46.3	45.3	44.2	43.2	42.2	41.1	40.1	39.1	38.2
56	55.7	55.4	54.9	54.3	53.6	52.9	52.0	51.1	50.1	49.1	48.0	47.0	45.9	44.8	43.7	42.7	41.6	40.6	39.6
58	57.7	57.4	56.9	56.3	55.6	54.7	53.9	52.9	51.9	50.8	49.7	48.6	47.5	46.4	45.3	44.2	43.1	42.0	41.0
60	59.7	59.3	58.8	58.2	57.5	56.6	55.7	54.7	53.7	52.6	51.4	50.3	49.1	48.0	46.9	45.7	44.6	43.5	42.4
62	61.7	61.3	60.8	60.1	59.4	58.5	57.6	56.5	55.5	54.3	53.2	52.0	50.8	49.6	48.4	47.2	46.1	44.9	43.8
64	63.7	63.3	62.8	62.1	61.3	60.4	59.4	58.4	57.2	56.1	54.9	53.7	52.4	51.2	50.0	48.8	47.6	46.4	45.3
66	65.7	65.3	64.7	64.0	63.2	62.3	61.3	60.2	59.0	57.8	56.6	55.3	54.1	52.8	51.5	50.3	49.1	47.8	46.7
68	67.7	67.2	66.7	66.0	65.1	64.2	63.1	62.0	60.8	59.6	58.3	57.0	55.7	54.4	53.1	51.8	50.5	49.3	48.1
70	69.7	69.2	68.6	67.9	67.0	66.1	65.0	63.8	62.6	61.3	60.0	58.7	57.3	56.0	54.7	53.3	52.0	50.7	49.5
72	71.6	71.2	70.6	69.9	69.0	68.0	66.9	65.7	64.4	63.1	61.7	60.4	59.0	57.6	56.2	54.9	53.5	52.2	50.9
74	73.6	73.2	72.6	71.8	70.9	69.8	68.7	67.5	66.2	64.8	63.5	62.0	60.6	59.2	57.8	56.4	55.0	53.6	52.3
76	75.6	75.2	74.5	73.7	72.8	71.7	70.6	69.3	68.0	66.6	65.2	63.7	62.3	60.8	59.3	57.9	56.5	55.1	53.7
78	77.6	77.1	76.5	75.7	74.7	73.6	72.4	71.1	69.8	68.3	66.9	65.4	63.9	62.4	60.9	59.4	58.0	56.5	55.2
80	79.6	79.1	78.4	77.6	76.6	75.5	74.3	73.0	71.6	70.1	68.6	67.1	65.5	64.0	62.5	61.0	59.5	58.0	56.6
82	81.6	81.1	80.4	79.6	78.5	77.4	76.1	74.8	73.3	71.9	70.3	68.8	67.2	65.6	64.0	62.5	61.0	59.4	58.0
84	83.6	83.1	82.4	81.5	80.5	79.3	78.0	76.6	75.1	73.6	72.0	70.4	68.8	67.2	65.6	64.0	62.4	60.9	59.4
86	85.6	85.0	84.3	83.4	82.4	81.2	79.9	78.4	76.9	75.4	73.7	72.1	70.5	68.8	67.2	65.5	63.9	62.3	60.8
88	87.6	87.0	86.3	85.4	84.3	83.1	81.7	80.2	78.7	77.1	75.5	73.8	72.1	70.4	68.7	67.1	65.4	63.8	62.2
90	89.6	89.0	88.3	87.3	86.2	85.0	83.6	82.1	80.5	78.9	77.2	75.5	73.7	72.0	70.3	68.6	66.9	65.2	63.6
92	91.5	91.0	90.2	89.3	88.1	86.8	85.4	83.9	82.3	80.6	78.9	77.1	75.4	73.6	71.8	70.1	68.4	66.7	65.1
94	93.5	93.0	92.2	91.2	90.0	88.7	87.3	85.7	84.1	82.4	80.6	78.8	77.0	75.2	73.4	71.6	69.9	68.1	66.5
96	95.5	94.9	94.1	93.1	92.0	90.6	89.1	87.5	85.9	84.1	82.3	80.5	78.6	76.8	75.0	73.1	71.4	69.6	67.9
98	97.5	96.9	96.1	95.1	93.9	92.5	91.0	89.4	87.7	85.9	84.0	82.2	80.3	78.4	76.5	74.7	72.8	71.0	69.3
100	99.5	98.9	98.1	97.0	95.8	94.4	92.8	91.2	89.4	87.6	85.7	83.8	81.9	80.0	78.1	76.2	74.3	72.5	70.7

TABLE No. 2.—*Conversion of slope distances to horizontal distances*

[Topographic Abney and 1-chain (100 links) tape]

Slope distance in links	Topographic															
	5	10	15	20	25	30	35	40	45	50	55	60	65	70	75	80
2	2.0	2.0	2.0	1.9	1.9	1.8	1.8	1.7	1.7	1.6	1.5	1.5	1.4	1.4	1.3	1.3
4	4.0	4.0	4.0	3.8	3.7	3.6	3.5	3.4	3.3	3.2	3.1	3.0	2.8	2.7	2.6	2.5
6	6.0	5.9	5.9	5.7	5.6	5.5	5.3	5.1	5.0	4.8	4.6	4.4	4.3	4.1	4.0	3.8
8	8.0	7.9	7.8	7.7	7.5	7.3	7.1	6.8	6.6	6.4	6.1	5.9	5.7	5.5	5.3	5.1
10	10.0	9.9	9.8	9.6	9.4	9.1	8.8	8.6	8.3	8.0	7.7	7.4	7.1	6.9	6.6	6.4
12	12.0	11.9	11.7	11.5	11.2	10.9	10.6	10.3	9.9	9.6	9.2	8.9	8.5	8.2	7.9	7.6
14	14.0	13.8	13.7	13.4	13.1	12.7	12.4	12.0	11.6	11.2	10.8	10.4	10.0	9.6	9.2	8.9
16	16.0	15.8	15.6	15.3	15.0	14.6	14.1	13.7	13.2	12.8	12.3	11.8	11.4	11.0	10.6	10.2
18	17.9	17.8	17.6	17.2	16.8	16.4	15.9	15.4	14.9	14.3	13.8	13.3	12.8	12.3	11.9	11.5
20	19.9	19.8	19.5	19.1	18.7	18.2	17.7	17.1	16.5	15.9	15.4	14.8	14.2	13.7	13.2	12.7
22	21.9	21.8	21.5	21.0	20.6	20.0	19.4	18.8	18.2	17.5	16.9	16.3	15.7	15.1	14.5	14.0
24	23.9	23.7	23.4	23.0	22.4	21.8	21.2	20.5	19.8	19.1	18.4	17.8	17.1	16.5	15.9	15.3
26	25.9	25.7	25.4	24.9	24.3	23.7	23.0	22.2	21.5	20.7	20.0	19.2	18.5	17.8	17.2	16.5
28	27.9	27.7	27.3	26.8	26.2	25.5	24.7	23.9	23.1	22.3	21.5	20.7	19.9	19.2	18.5	17.8
30	29.9	29.7	29.3	28.7	28.1	27.3	26.5	25.7	24.8	23.9	23.0	22.2	21.4	20.6	19.8	19.1
32	31.9	31.6	31.2	30.6	29.9	29.1	28.3	27.4	26.4	25.5	24.6	23.7	22.8	22.0	21.1	20.4
34	33.9	33.6	33.2	32.5	31.8	31.0	30.0	29.1	28.1	27.1	26.1	25.2	24.2	23.3	22.5	21.6
36	35.9	35.6	35.1	34.5	33.7	32.8	31.8	30.8	29.7	28.7	27.7	26.6	25.6	24.7	23.8	22.9
38	37.9	37.6	37.1	36.4	35.5	34.6	33.6	32.5	31.4	30.3	29.2	28.1	27.1	26.1	25.1	24.2
40	39.9	39.5	39.0	38.3	37.4	36.4	35.3	34.2	33.0	31.9	30.7	29.6	28.5	27.4	26.4	25.5
42	41.9	41.5	41.0	40.2	39.3	38.2	37.0	35.9	34.7	33.5	32.3	31.1	29.9	28.8	27.7	26.7
44	43.9	43.5	42.9	42.1	41.1	40.1	38.9	37.6	36.4	35.1	33.8	32.6	31.3	30.2	29.1	28.0
46	45.9	45.5	44.9	44.0	43.0	41.9	40.6	39.3	38.0	36.7	35.3	34.0	32.8	31.6	30.4	29.3
48	47.9	47.5	46.8	45.9	44.9	43.7	42.4	41.0	39.7	38.3	36.9	35.3	34.2	32.9	31.7	30.5
50	49.9	49.4	48.8	47.9	46.8	45.5	44.2	42.8	41.3	39.9	38.4	37.0	35.6	34.3	33.0	31.8
52	51.9	51.4	50.7	49.8	48.6	47.3	45.9	44.5	43.0	41.4	39.9	38.5	37.0	35.7	34.4	33.1
54	53.8	53.4	52.7	51.7	50.5	49.2	47.7	46.2	44.6	43.0	41.5	40.0	38.5	37.0	35.7	34.4
56	55.8	55.4	54.6	53.6	52.4	51.0	49.5	47.9	46.3	44.6	43.0	41.4	39.9	38.4	37.0	35.6
58	57.8	57.3	56.6	55.5	54.2	52.8	51.2	49.6	47.9	46.2	44.6	42.9	41.3	39.8	38.3	36.9
60	59.8	59.3	58.5	57.4	56.1	54.6	53.0	51.3	49.6	47.8	46.1	44.4	42.7	41.2	39.6	38.2
62	61.8	61.3	60.5	59.3	58.0	56.4	54.8	53.0	51.2	49.4	47.6	45.9	44.2	42.5	41.0	39.5
64	63.8	63.3	62.4	61.2	59.9	58.3	56.5	54.7	52.9	51.0	49.2	47.4	45.6	43.9	42.3	40.7
66	65.8	65.3	64.4	63.2	61.7	60.1	58.3	56.4	54.5	52.6	50.7	48.8	47.0	45.3	43.6	42.0
68	67.8	67.2	66.3	65.1	63.6	61.9	60.1	58.1	56.2	54.2	52.2	50.3	48.4	46.4	44.9	43.3
70	69.8	69.2	68.3	67.0	65.5	63.7	61.8	59.9	57.8	55.8	53.8	51.8	49.9	48.0	46.2	44.5
72	71.8	71.2	70.2	68.9	67.3	65.5	63.6	61.6	59.5	57.4	55.3	53.3	51.3	49.4	47.6	45.8
74	73.8	73.2	72.2	70.8	69.2	67.4	65.4	63.3	61.1	59.0	56.8	54.8	52.7	50.8	48.9	47.1
76	75.8	75.1	74.1	72.7	71.1	69.2	67.1	65.0	62.8	60.6	58.4	56.2	54.1	52.1	50.2	48.4
78	77.8	77.1	76.1	74.6	72.9	71.0	68.9	66.7	64.4	62.2	59.9	57.7	55.6	53.5	51.5	49.6
80	79.8	79.1	78.0	76.6	74.8	72.8	70.7	68.4	66.1	63.8	61.5	59.2	57.0	54.9	52.9	50.9
82	81.8	81.1	80.0	78.5	76.6	74.6	72.4	70.1	67.8	65.4	63.0	60.7	58.4	56.3	54.2	52.2
84	83.8	83.1	81.9	80.4	78.6	76.5	74.2	71.8	69.4	67.0	64.5	62.2	59.8	57.6	55.5	53.5
86	85.8	85.0	83.9	82.3	80.4	78.3	76.0	73.5	71.1	68.5	66.1	63.6	61.3	59.0	56.8	54.7
88	87.7	87.0	85.8	84.2	82.3	80.1	77.7	75.3	72.7	70.1	67.6	65.1	62.7	60.4	58.1	56.0
90	89.7	89.0	87.8	86.1	84.2	81.9	79.5	77.0	74.4	71.7	69.1	66.6	64.1	61.7	59.5	57.3
92	91.7	91.0	89.7	88.0	86.0	83.8	81.3	78.7	76.0	73.3	70.7	68.1	65.5	63.1	60.8	58.5
94	93.7	92.9	91.7	90.0	87.9	85.6	83.0	80.4	77.7	74.9	72.2	69.6	67.0	64.7	62.1	59.8
96	95.7	94.9	93.6	91.9	89.8	87.4	84.8	82.1	79.3	76.5	73.7	71.0	68.4	65.9	63.4	61.1
98	97.7	96.9	95.6	93.8	91.6	89.2	86.6	83.8	81.0	78.1	75.3	72.5	69.8	67.2	64.7	62.4
100	99.7	98.9	97.5	95.7	93.5	91.0	88.3	85.5	82.6	79.7	76.8	74.0	71.2	68.6	66.1	63.6

The difference in elevation is obtained by multiplying the horizontal distance in chains (not in links) by the slope (Abney reading).

Equivalents of percents in degrees

Percent	Degrees	Percent	Degrees	Percent	Degrees	Percent	Degrees
	° '		° '		° '		° '
1	34	26	14 34	51	27 01	76	37 14
2	1 09	27	15 07	52	27 28	77	37 36
3	1 43	28	15 39	53	27 55	78	37 57
4	2 17	29	16 10	54	28 22	79	38 19
5	2 52	30	16 42	55	28 49	80	38 40
6	3 26	31	17 13	56	29 15	81	39 00
7	4 00	32	17 45	57	29 41	82	39 21
8	4 34	33	18 16	58	30 07	83	39 42
9	5 09	34	18 47	59	30 32	84	40 02
10	5 43	35	19 17	60	30 58	85	40 22
11	6 17	36	19 48	61	31 23	86	40 42
12	6 51	37	20 18	62	31 48	87	41 01
13	7 24	38	20 48	63	32 13	88	41 21
14	7 58	39	21 18	64	32 37	89	41 40
15	8 32	40	21 48	65	33 01	90	41 59
16	9 05	41	22 18	66	33 25	91	42 18
17	9 39	42	22 47	67	33 49	92	42 37
18	10 12	43	23 16	68	34 13	93	42 55
19	10 45	44	23 45	69	34 36	94	43 14
20	11 19	45	24 14	70	35 00	95	43 32
21	11 52	46	24 42	71	35 22	96	43 50
22	12 24	47	25 10	72	35 45	97	44 08
23	12 57	48	25 38	73	36 08	98	44 25
24	13 30	49	26 06	74	36 30	99	44 43
25	14 02	50	26 34	75	36 52	100	45 00

Equivalents of degrees in percent

Degrees	Percent	Degrees	Percent	Degrees	Percent	Degrees	Percent
1	1.74	16	28.67	31	60.09	46	103.55
2	3.49	17	30.57	32	62.49	47	107.24
3	5.24	18	32.49	33	64.94	48	111.06
4	6.99	19	34.43	34	67.45	49	115.04
5	8.75	20	36.40	35	70.02	50	119.18
6	10.51	21	38.39	36	72.65	51	123.49
7	12.28	22	40.40	37	75.35	52	127.99
8	14.05	23	42.45	38	78.13	53	132.70
9	15.84	24	44.52	39	80.98	54	137.64
10	17.63	25	46.63	40	83.91	55	142.81
11	19.44	26	48.77	41	86.93	56	148.26
12	21.26	27	50.95	42	90.04	57	153.99
13	23.09	28	53.17	43	93.25	58	160.03
14	24.93	29	55.43	44	96.57	59	166.43
15	26.80	30	57.73	45	100.00	60	173.20

Table 3 is for the purpose of converting percent to degree graduations and conversely for converting degree to percent graduations.

TABLE No. 4

Equivalents of percents in topographic graduations

Percent	Topographic	Percent	Topographic	Percent	Topographic	Percent	Topographic
1	0.66	26	17.16	51	33.66	76	50.16
2	1.32	27	17.82	52	34.32	77	50.82
3	1.98	28	18.48	53	34.98	78	51.48
4	2.64	29	19.14	54	35.64	79	52.14
5	3.30	30	19.80	55	36.30	80	52.80
6	3.96	31	20.46	56	36.96	81	53.46
7	4.62	32	21.12	57	37.62	82	54.12
8	5.28	33	21.78	58	38.28	83	54.78
9	5.94	34	22.44	59	38.94	84	55.44
10	6.60	35	23.10	60	39.60	85	56.10
11	7.26	36	23.76	61	40.26	86	56.76
12	7.92	37	24.42	62	40.92	87	57.42
13	8.58	38	25.08	63	41.58	88	58.08
14	9.24	39	25.74	64	42.24	89	58.74
15	9.90	40	26.40	65	42.90	90	59.40
16	10.56	41	27.06	66	43.56	91	60.06
17	11.22	42	27.72	67	44.22	92	60.72
18	11.88	43	28.38	68	44.88	93	61.38
19	12.54	44	29.04	69	45.54	94	62.04
20	13.20	45	29.70	70	46.20	95	62.70
21	13.86	46	30.36	71	46.86	96	63.36
22	14.52	47	31.02	72	47.52	97	64.02
23	15.18	48	31.68	73	48.18	98	64.68
24	15.84	49	32.34	74	48.84	99	65.34
25	16.50	50	33.00	75	49.50	100	66.00

Equivalents of topographic graduations in percents

Topographic	Percent	Topographic	Percent	Topographic	Percent	Topographic	Percent
1	1.51	21	31.81	41	62.11	61	92.41
2	3.03	22	33.33	42	63.63	62	93.93
3	4.54	23	34.84	43	65.14	63	95.44
4	6.06	24	36.36	44	66.66	64	96.96
5	7.57	25	37.87	45	68.17	65	98.47
6	9.09	26	39.39	46	69.69	66	99.99
7	10.60	27	40.90	47	71.20	67	101.50
8	12.12	28	42.42	48	72.72	68	103.02
9	13.63	29	43.93	49	74.23	69	104.53
10	15.15	30	45.45	50	75.75	70	106.05
11	16.67	31	46.97	51	77.27	71	107.57
12	18.18	32	48.48	52	78.78	72	109.08
13	19.70	33	50.00	53	80.30	73	110.60
14	21.21	34	51.51	54	81.81	74	112.11
15	22.73	35	53.03	55	83.33	75	113.63
16	24.21	36	54.54	56	84.84	76	115.14
17	25.76	37	56.66	57	86.36	77	116.66
18	27.27	38	57.57	58	87.87	78	118.17
19	28.79	39	59.09	59	89.39	79	119.69
20	30.30	40	60.60	60	90.90	80	121.20

Table 4 furnishes equivalents between percent and topographic graduations.

The 66-foot base from which topographic graduations are derived is approximately two-thirds of the 100-feet base from which the percent graduations are derived. The following rule which is developed from the proportional relation that exists between the 66 to 100 foot bases will be found convenient where no great amount of accuracy is desired: Multiply topographic Abney readings by 1 ½ to convert them to per cent readings; multiply percent readings by 2/3 to convert them to topographic readings. From table 4, a percent reading of 100 is equivalent to a topographic reading of 66. From the rule 100 x 2/3=66 2/3, therefore the amount of error obtained by using the rule is two-thirds percent for a reading of 100 percent and one-third percent for a reading of 50 percent.

TABLE No. 5

Equivalents of degrees in topographic graduations

Degrees	Topographic	Degrees	Topographic	Degrees	Topographic	Degrees	Topographic
1	1.15	16	18.93	31	39.66	46	68.34
2	2.30	17	20.18	32	41.24	47	70.78
3	3.46	18	21.44	33	42.86	48	73.30
4	4.62	19	22.73	34	44.52	49	75.92
5	5.77	20	24.02	35	46.21	50	78.66
6	6.94	21	25.33	36	47.95	51	81.50
7	8.10	22	26.67	37	49.73	52	84.48
8	9.28	23	28.02	38	51.57	53	87.58
9	10.45	24	29.39	39	53.45	54	90.84
10	11.64	25	30.78	40	55.38	55	94.26
11	12.83	26	32.19	41	57.37	56	97.85
12	14.03	27	33.63	42	59.43	57	101.63
13	15.24	28	35.09	43	61.55	58	105.62
14	16.46	29	36.58	44	63.74	59	109.84
15	17.68	30	38.11	45	66.00	60	114.32

Equivalents of topographic graduations in degrees

Topographic	Degrees		Topographic	Degrees		Topographic	Degrees		Topographic	Degrees	
	°	′		°	′		°	′		°	′
1	0	52	21	17	39	41	31	51	61	42	45
2	1	44	22	18	26	42	32	28	62	43	13
3	2	36	23	19	13	43	33	05	63	43	40
4	3	28	24	19	59	44	33	41	64	44	07
5	4	20	25	20	45	45	34	17	65	44	34
6	5	12	26	21	30	46	34	53	66	45	00
7	6	04	27	22	15	47	35	27	67	45	26
8	6	55	28	22	59	48	36	02	68	45	51
9	7	46	29	23	43	49	36	35	69	46	16
10	8	37	30	24	27	50	37	09	70	46	41
11	9	28	31	25	10	51	37	42	71	47	05
12	10	18	32	25	52	52	38	14	72	47	29
13	11	08	33	26	34	53	38	46	73	47	53
14	11	59	34	27	15	54	39	17	74	48	16
15	12	48	35	27	56	55	39	48	75	48	39
16	13	38	36	28	37	56	40	19	76	49	02
17	14	27	37	29	17	57	40	49	77	49	24
18	15	15	38	29	56	58	41	19	78	49	46
19	16	04	39	30	35	59	41	48	79	50	07
20	16	51	40	31	13	60	42	16	80	50	29

Table 5 is for the purpose of converting degree to topographic graduations and, conversely, for converting topographic to degree graduations.

[1] A number of old-style (not band steel) 100-link chains are still in use in some localities.

46

CPSIA information can be obtained
at www.ICGtesting.com
Printed in the USA
LVHW041554210621
690768LV00004B/900

9 781789 870503